Transport o

A PICTORIAL CELEBRATION OF THE WEST SOMERSET RAILWAY

Cedric Dunmall

THE EXMOOR PRESS

First published 1990
by THE EXMOOR PRESS
Dulverton, TA22 9EX

BRITISH LIBRARY CATALOGUING IN PUBLICATION DATA

Dunmall, Cedric

Transport of Delight: a pictorial celebration of the West Somerset Railway.
1. Somerset. West Somerset (District) Railway Services. West Somerset Railway.
385.0942385

ISBN 0-900131-64-0

Front Cover: Pannier tank No. 6412 makes light work of Crowcombe Bank with an early season train from Minehead to Bishops Lydeard.

Back Cover: Sunset Special at Blue Anchor Bay with Evening Star.

Printed in Great Britain by
Williton Printers, Somerset

CONTENTS

		Page
	"How I Fell in Love with Steam."	4
	Introduction — the Story of the West Somerset Railway.	6
1.	Down the Branch from Taunton To Minehead. Stations — Signal Boxes — Signals.	8
2.	The Hidden Railway. The Permanent Way.	24
3.	Motive Power. Steam and Diesel "On Shed" and "On the Road."	35
4.	Rolling Stock. Carriages and Diesel Multiple Units.	52
5.	The Scenic Railway. With Steam and Diesel.	63
6.	Events. Films — Charter Trains — Galas — Rallies.	78
7.	EVENING STAR. Special Visitor and Star of the Show.	89

The Author

Cedric Dunmall was born at Dartford, Kent, in 1937. A passion for railways and an interest in photography were combined when he became a volunteer fireman and unofficial photographer with the West Somerset Railway. This unique perspective has led to a remarkable collection of pictures celebrating the life of Britain's longest preserved line.

"HOW I FELL IN LOVE WITH STEAM"

A Southern tank engine with two smartly turned-out green coaches working the branch line from Farningham Road to Gravesend West Street must be partly responsible for this, my selection of photographs of the West Somerset Railway.

Perhaps a bold statement, but that particular corner of Kent was the origin of my love of the steam train and a determination to discover more in later years. The sheer delight, on a summer Saturday evening in the late '40s, at Gravesend's deserted platform, when the friendly engine crew asked if I 'wanted to come up' was an event never to be forgotten. Newspaper was spread at the rear of the cab to protect a small lad's clean clothes — we were to run chimney first — and I was lifted aboard. Many years back it may be, but I can recall that the fireman had a pretty easy time over those seven miles of branch line, compared with the 'turn' over the West Somerset, now related from experience.

The starting signal was raised: we were on our way, clanking through the steep chalk cuttings so typical of north Kent, past a dim and long abandoned Rosherville Halt, over the A2 road and drifting to a stop at Southfleet's island platform. Of course there were no passengers — there seldom were, except for school trains and the Saturday shoppers. Regulator open and away along straight track making for the all-timber construction Longfield Halt. The big event here, observed from the last train of the day, was the guard's climb to the top of wooden steps, leading to a quiet country lane above, in order to extinguish the solitary lamp. This task expeditiously performed, with a green light and a short whistle we were off for the sprint to Pinden Junction on the main Kent coast line to London Victoria. Passing the junction semaphore, and with a whistle for the signal man in his lonely box, engine rocking wildly over the crossings and accelerating down the gradient, we were soon out of more steep chalk cuttings and high over the Darent Valley on its viaduct, past my house, never before seen from this angle, and into Farningham Road station. It was all over. I can remember Father appearing beside the engine and lifting me down . . .

I say 'partly responsible', for it was not until secondary school days in the 50s and active membership of its photographic club, that I became the proud owner of pocket money cameras in the form of Kodak's 'Baby Brownie' and the 'Coronet Box'. Unite steam and camera, in a home shaken regularly to its foundations by express trains with names such as 'Golden Arrow' and 'Thanet Belle' as they raced over Sole Street Bank, and you have a fine recipe for recording the preservation scene to follow.

Strong association with the old 'Southern' continued for me into the '60s accompanied, at one stage, by an R.A.F. uniform between Waterloo and Basingstoke. In due course, when electrics and diesels reigned supreme on home territory, came a move to the West Country and to within a mile or so of the Culm Valley railway, with its milk trains working back and forth between Hemyock Creamery and Tiverton Junction.

Soon was to come news of a scheme to reopen the railway between Taunton and Minehead and thus, with portable generator and vacuum cleaner to hand, my link with the railway began in Taunton's West Yard — cleaning the first two coaches to be used on the preserved West Somerset Railway. In that same West Yard at the time stood Bagnall saddle tank locos, VICTOR and VULCAN. In steamable condition by late 1975 VICTOR was soon to be moved to the West Somerset Railway at Norton Fitzwarren; and it was there, one mild December morning, that my introduction to firing was made. The rest, as they say, is history . . .

CEDRIC DUNMALL

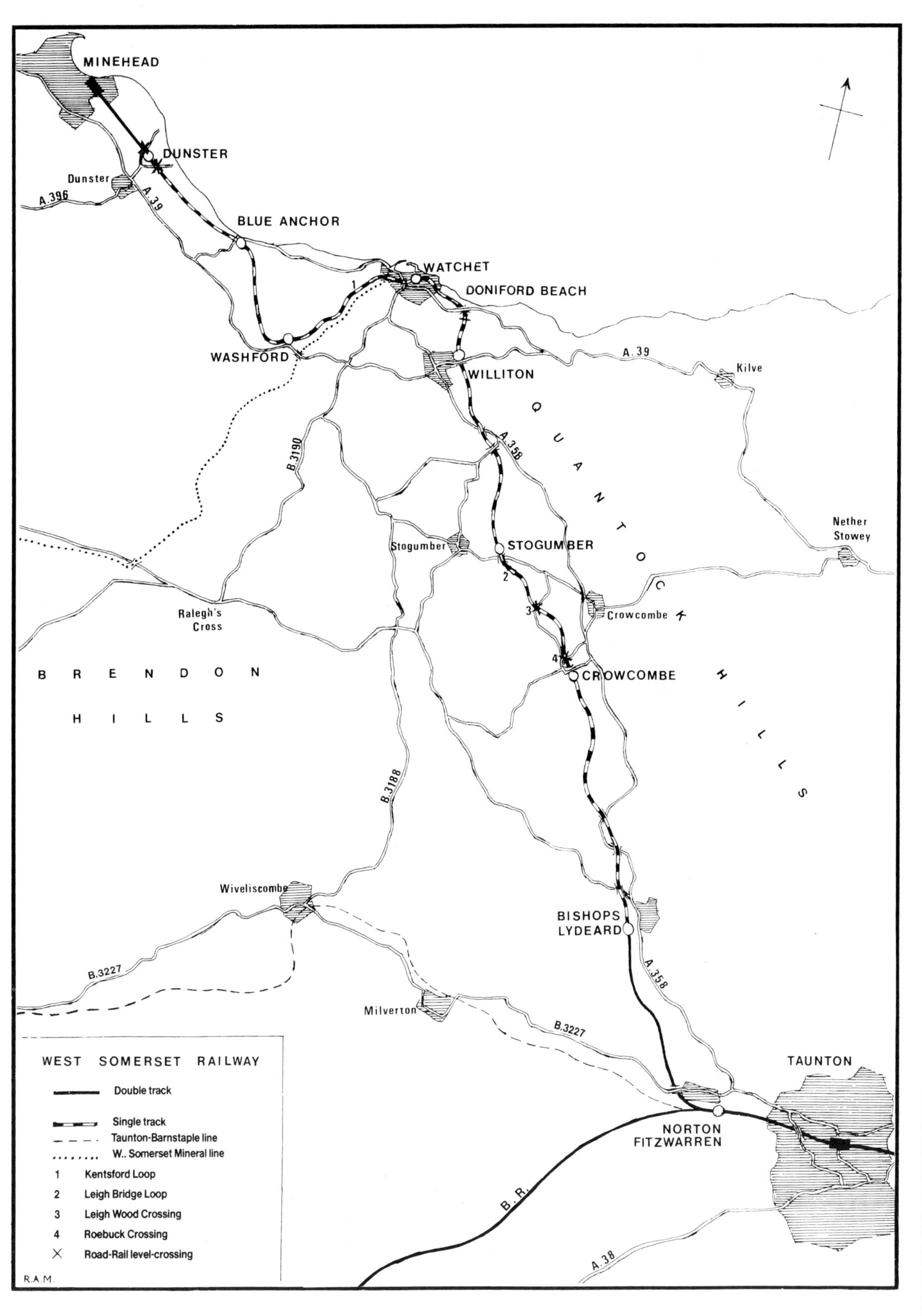
MINEHEAD
DUNSTER
Dunster
A.396
A.39
BLUE ANCHOR
WATCHET
DONIFORD BEACH
WASHFORD
WILLITON
A.39
Kilve
QUANTOCK HILLS
A.358
B.3190
Nether
Stowey
Stogumber
STOGUMBER
Crowcombe
Ralegh's
Cross
CROWCOMBE
BRENDON HILLS
B.3188
Wiveliscombe
BISHOPS
LYDEARD
A.358
B.3227
Milverton
B.3227
TAUNTON
NORTON
FITZWARREN
B.R.
A.38
WEST SOMERSET RAILWAY
Double track
Single track
Taunton-Barnstaple line
W.. Somerset Mineral line
1 Kentsford Loop
2 Leigh Bridge Loop
3 Leigh Wood Crossing
4 Roebuck Crossing
Road-Rail level-crossing
R.A.M.

INTRODUCTION The Story of the West Somerset Railway

The West Somerset Railway, subject of this personal photographic collection, was built in two stages. The first, authorised in 1857, was eventually completed — after constructional and financial difficulties — in 1862, and ran from a junction with the Bristol and Exeter Railway at Norton Fitzwarren (2 miles west of Taunton) to Watchet, a distance of some 14 miles. The second stage, known as the Minehead Railway, was not completed until 1874, a distance of about 8¾ miles. The total distance was just under 23 miles. Both sections of the line were worked first by the Bristol and Exeter Railway, then by the Great Western Railway, and both were converted to standard narrow gauge (4ft. 8½ ins.) in 1882.

Traffic for the first thirty years or so was not heavy, with four passenger trains each way initially between Taunton and Watchet and later between Taunton and Minehead on weekdays. However, it slowly improved with the growth of Watchet as a port, thanks to the import of coal from South Wales, and of esparto grass and wood pulp for the paper mills at Watchet and at Silverton in Devon. For a few years iron ore from the Brendon Hills was shipped to Newport and on to Ebbw Vale, but ceased after the closure of the mines in 1883. There was never any permanent connection with the West Somerset Mineral Railway, though both companies ran lines into Watchet harbour.

Passenger traffic increased with the development of Minehead as a holiday and seaside resort. Longer platforms were built at most of the intermediate stations, together with crossing loops and other facilities. By 1934 there was a summer service of ten trains each way on weekdays and five on Sundays, with provision for through coaches to Paddington. The average running time between Taunton and Minehead was 50 minutes. At Watchet the peak year for goods was 1937, when nearly 60,000 tons were forwarded and received, over 15,000 parcels despatched, and just under 23,000 tickets issued, including 214 seasons. At Minehead the peak year was 1913 when 48,578 tickets were sold. Thereafter the turnover declined though both kinds of traffic — passengers and goods — remained brisk right up to 1939, and the line supplied an essential service during both world wars.

Under the Transport Act 1947, the branch was nationalised, soon followed by a series of economies that depleted the condition of the track and the stations, closed signal boxes, and reduced staff (from 50 in 1938) to 11. All freight services were withdrawn in 1966. It was a lamentable process, and morale fell to zero when, in 1963, British Railways issued a report

Bristol and Exeter Railway 4-4-0 saddletank engine No. 68, drawing one of the first trains into Watchet, c. 1867. *(James Date, courtesy R. J. Sellick Collection.)*

which proposed the closure of the line in its entirety. Following some difficulty in providing alternative bus services, this duly took place on 4th January, 1971. By that date however steps were already being taken to purchase the property from the British Railways Board and restore some kind of passenger service. The West Somerset Railway Co. Ltd. (the original name) was incorporated as a private company on 5th May, 1971, only four months after closure of the line; and slowly, though painfully, the necessary funds were assembled and procedures followed towards restoration. After a public inquiry in 1974, a Light Railway Order was obtained and the line transferred to Somerset County Council in the following year. Having secured a 20-year lease, the Company 'went public' on 31 March, 1976, and issued an initial shareholding of £65,000, a sum later increased in a series of new issues as fresh finance was required. The most recent issue, launched in November, 1988, has enabled the Company to buy back the lease from Somerset County Council and start a heavy programme of investment in fixed and other equipment. This has coincided with a marked improvement in annual trading. Although it has not yet proved possible to extend services beyond the effective terminus at Bishops Lydeard to link with BR at Taunton Station, passengers carried in 1989 exceeded 100,000 (up over 30% on any previous year), while retail business at the shop on Minehead Station was also a record.

None of this progress would have been possible without the aid of volunteers, organised in 1971 as the West Somerset Railway Association with the sole purpose of assisting the Company in all departments of line operation —providing (in addition to the paid staff) engine drivers and firemen, guards, signalmen, station and permanent way staff, mechanics and many others, including a team that mans the QUANTOCK BELLE dining car train on occasional Saturdays and Sundays. The Association now has nearly 2,000 members, of whom perhaps 100 are actively engaged in the practical working of the line.

The West Somerset Railway is the longest preserved line in Britain. Its locomotives, coaches, stations, signalling, and all the variety of line operation are set out and illustrated in the following pages. They are the essential components of a remarkable success story, and the foundations of an achievement fully worthy of celebration.

Williton, January 1971, British Rail's last day of working. *(Robin Madge.)*

45

Chapter 1

DOWN THE BRANCH

Above:
A tantalising glimpse of how it could be. Taunton Station, yet to be reached by regular WSR services. A crowded bay platform, and a Manor Class locomotive about to depart westward.

Left:
From the window of Silk Mills crossing signal box during a late 1970s winter. The up relief or Minehead line lies undisturbed beneath light snow. Much of the trackwork, extreme left, was recovered for further use on the WSR. The site today is unrecognisable with M.A.S. (modern signalling), and the crossing controlled from Exeter signalling centre by closed circuit television.

Above:
Taunton Cider Works at Norton Fitzwarren. The point at which the WSR branches away from BR's main line to the west and the site of the now demolished Norton Fitzwarren Station. Here we see a visiting special charter train from Minehead hauled by loco 7F No. 53808, and being shunted by Taunton Cider's road-rail vehicle.

Right:
Once covered with numerous tracks to Exeter, Barnstaple and Minehead, this is all that remains of Norton Fitzwarren junction. The WSR curves away to the right towards its first station at Bishops Lydeard.

Above:
Bishops Lydeard Station, the present south terminus of the West Somerset Railway. Inherited from BR without an awning to the main station building. In the background is the old station master's house, now a summer holiday 'let'.

Below:
The south end of Bishops Lydeard Station showing the near-restored signal box on the right. This station is the headquarters of the WSR Association, and much voluntary work radiates from here.

Opposite Top:
Out in the country, near Bishops Lydeard, Bagnall saddle tank 'Victor' and prairie No. 5572 drift past harvested fields.

Opposite Below:
Early season at Combe Florey. A train for Minehead climbs steadily for Crowcombe beneath the Quantock Hills.

Clouds rest over the summit at Crowcombe Station as EVENING STAR storms through.

Above:
Crowcombe Station, the highest on the railway and at the head of a long climb from Bishops Lydeard. Justifiably honoured with a First Class award from the Association of Railway Preservation Societies, the building, platforms and gardens are carefully maintained.

Right:
Crowcombe Station lies hidden in the vale between the Brendon Hills and the Quantocks, well off the beaten track with access to delightful walks in the countryside. Once known as Crowcombe Heathfield, the name deriving from the neighbourhood, it lies at some distance from its village — a common feature in old GWR branch line days.

Bright afternoon sunshine lights the passing of Collett loco. No. 3205 as it attacks Crowcombe Bank with a specially chartered QUANTOCK BELLE.

Autumn leaves cover Crowcombe Station's platform as 7F No. 53808 shatters the calm, heading through with a non-stop train from Minehead.

Right:
Stogumber Station, like Crowcombe, is a stiff walk away from the village. The down side waiting shelter, isolated by the removal of rotting platform timbers, has now been restored.

Below:
Williton Station, midway between Bishops Lydeard and Minehead, is a fully signalled crossing point for trains. It is also the centre for Diesel and Electric Group activities on the WSR. Bristol and Exeter Railway architecture is evident here, as is a relic of broad gauge days visible in the wide space between the tracks.

Left:
The approach to Watchet Station with goods shed, now privately used. Dockside cargoes seen here would once have been rail-hauled. Sidings once filled the right-hand foreground, leading to the harbour's edge. A pedestrian, light-controlled crossing is located at the near end of a sharply curved single platform.

Below:
Despite the name (Midford) displayed on the tiny signal cabin, this *really is* Washford Station, home of the Somerset and Dorset Trust. Their pride and joy, loco 7F No. 53808, stands in the extensive sidings. The main station building houses a fine museum in memory of the 'Serene and Delightful' railway that once crossed the Mendip Hills.

Opposite Above:
Summer service at Blue Anchor Station with pannier tank No. 6412 awaiting the arrival of EVENING STAR with a train from Minehead.

Opposite Below:
From the footplate, Dunster Station and goods shed.

Minehead Station and sea-front as seen from North Hill. The long straight stretch of line in the centre leads to Dunster in the distance.

Above:
Beneath the canopy of Minehead Station, headquarters of the WSR Company, are displayed a GW horse-drawn lorry, a withy carrier marked Bridgwater & Taunton Canal, Durston, and an old-type telephone box, dating from the 1920s.

Left:
A damp winter morning at Minehead as Collett No. 3205 heads along the run-round loop.

Minehead Station is the northern terminus of the line. Its twin-faced platform was lengthened in 1934 to accommodate holiday trains from many parts of the country. GWR pannier tank No. 6412 hurries a train away to Williston. North Hill rises behind the steam cloud.

2B 34

Chapter 2

THE HIDDEN RAILWAY

Left:
A scene from the early days of preservation on the railway, in which a train load of concrete sleepers is ready to move off from Bishops Lydeard down the line. The loco is taking on water by pump from a tank wagon.

Below:
Working in conjunction with the WSR Company gang, weekend volunteers are here seen changing sleepers at Bishops Lydeard. To prove that the ladies can do it, the team includes Angela — as well as Dave — Randall.

With shirt sleeves rolled up, Ivan Willicombe sets about bolting together lengths of point rodding. The short post to the left will support signal control wires.

Left:
Following the closure of Westbury South Signal Box on BR Western Region in the late 1970s, the lever frame was purchased by the WSR. Work on restoring one section for re-installation in the Bishops Lydeard signal box was soon under way. Here the levers await attention in the goods shed.

Right and Below:
Painting and attention to mechanical detail completed, the first lever is placed in position. Arthur and Ivan Willicombe, working as volunteers at evenings and weekends, see the project through to completion at Bishops Lydeard.

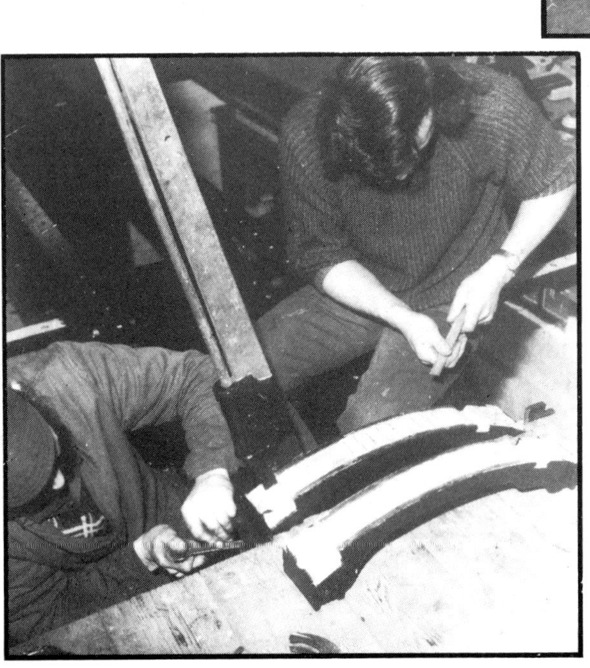

Above:
Bishops Lydeard Signal Box, re-decorated and complete with the lever frame installed.

Right:
Outside the Signal Box, the heavy work continues with digging into heavy clay to construct a 'lead-away', where point rodding and signal wires emerge from the base of the building.

Left:
Brute force and a bit more ensure a firm fixing for Bishops Lydeard's newly restored 'up side' name board in April 1989. Severe exposure to wind and weather at this spot means constant maintenance.

Below:
When you have no automatic carriage-washing plant, a bucket, hose and mop, plus elbow work, must do the job. Here a Park Royal DMU, in early maroon and cream livery, is undergoing the treatment at Bishops Lydeard during a lunch-time stopover.

Right:
Fifty feet long, Swindon-built, 'Siphon G' Milk Van is being painted by volunteers at Bishops Lydeard. A typical Saturday scene on the railway, where fresh help is always welcome.

Below:
The future Crowcombe Signal Box also requires urgent attention with the paint brush — in the hands of Walter Harris, the stationmaster. Transported by road from South Wales, it will eventually be moved to the opposite platform, be fitted out, and then control the passing loop at Crowcombe Station. Behind can be seen a privately owned camping coach.

Right:
Between trains at Blue Anchor Station, the duty signalman - cum - booking clerk makes a start on repainting the platform edge white line.

Below:
Visitors to Blue Anchor Station Museum are probably unaware of the activity on the awning roof overhead. Repairs and painting of roof sheets are necessary to make it proof against the elements.

On a rare day in late summer, when no passenger trains are running from Bishops Lydeard, essential maintenance is attended to along the line. Here is 7F No.53808, coupled up to an ex-GWR 5-ton hand crane and matchtruck, converted from an open wagon.

Right:
A winter's day near Combe Florey finds our permanent way train working hard on the four-mile climb to Crowcombe.

Below:
Having worked stock from Minehead, newly rebuilt prairie tank No. 4561 spends the remainder of a dull grey afternoon on 10th November, 1989, at Bishops Lydeard, shunting the yard and assisting the permanent way department.

Chapter 3

MOTIVE POWER

Left:
The source of steam power on the WSR for many years. Bagnall saddle tanks, VICTOR and VULCAN, take on water at Bishops Lydeard — in this instance from a rail tank wagon. To the left is a Ruston & Hornsby diesel shunter originally in service with United Dairies at Chard Junction.

Below:
VULCAN, No. 2994, in apple-green livery sponsored by Castrol, stands at Bishops Lydeard at the head of a Minehead train.

A running inspection by fireman R. S. Hindley, with Don Haynes at the controls of VICTOR, No. 2996. Both men have been mines of information for the author while working on the locomotives of the WSR.

Overleaf Above:
Prairie tank No. 5572 receives 'on the road' attention from its driver whilst a brief stop is made at Crowcombe Station.

Overleaf Below:
From the trackside. In Gala Week 1989, 7F 53808 has become 53807 to celebrate the last run of the class over the Somerset and Dorset's Mendip Hills 25 years ago.

The wooded setting of Crowcombe surrounds 7F No. 53808 as it climbs into the station with a train from Bishops Lydeard.

Left:
August 1987 finds a sparkling 7F No. 53808 making steam for its first trial working with two coaches to Bishops Lydeard. The ex-S & DJR loco's restoration was complete — 17 years after removal from Barry Scrapyard.

Below:
The trial run under way with a pause at Washford Station. This site has been greatly improved, including the erection of a new, twin-road, shed in 1988.

Above:
S & DJR loco 7F No. 53808 is seen through the cab window of VICTOR.

Below:
The headboard reads 'Wedding Express' in readiness for a reception charter train to be hauled by 7F No. 53808 between Bishops Lydeard and Minehead and back.

Overleaf Above:
Steam meets Diesel at Minehead on a summer Saturday. EVENING STAR standing on the shed road.

Overleaf Below:
From the trackside. Hymek D7017 in immaculate condition at Blue Anchor Station.

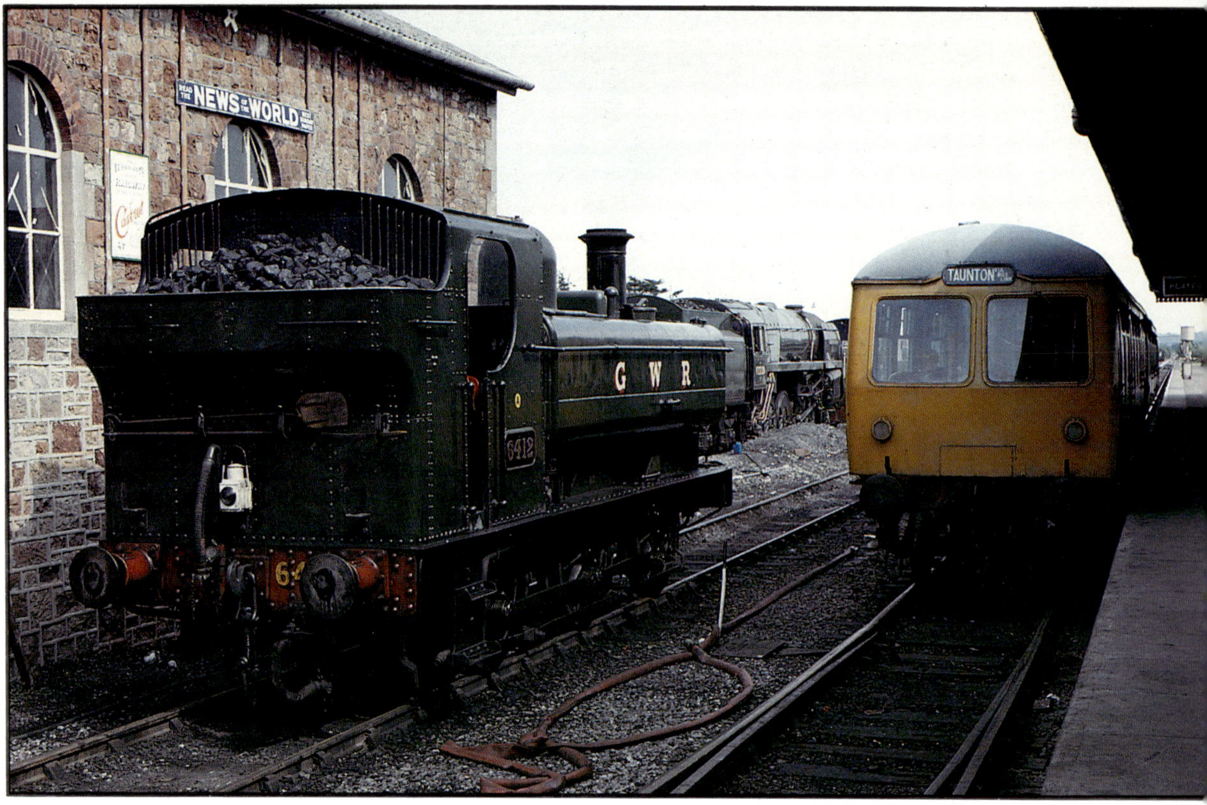
NEWS OF THE WORLD
G W R
6412
TAUNTON

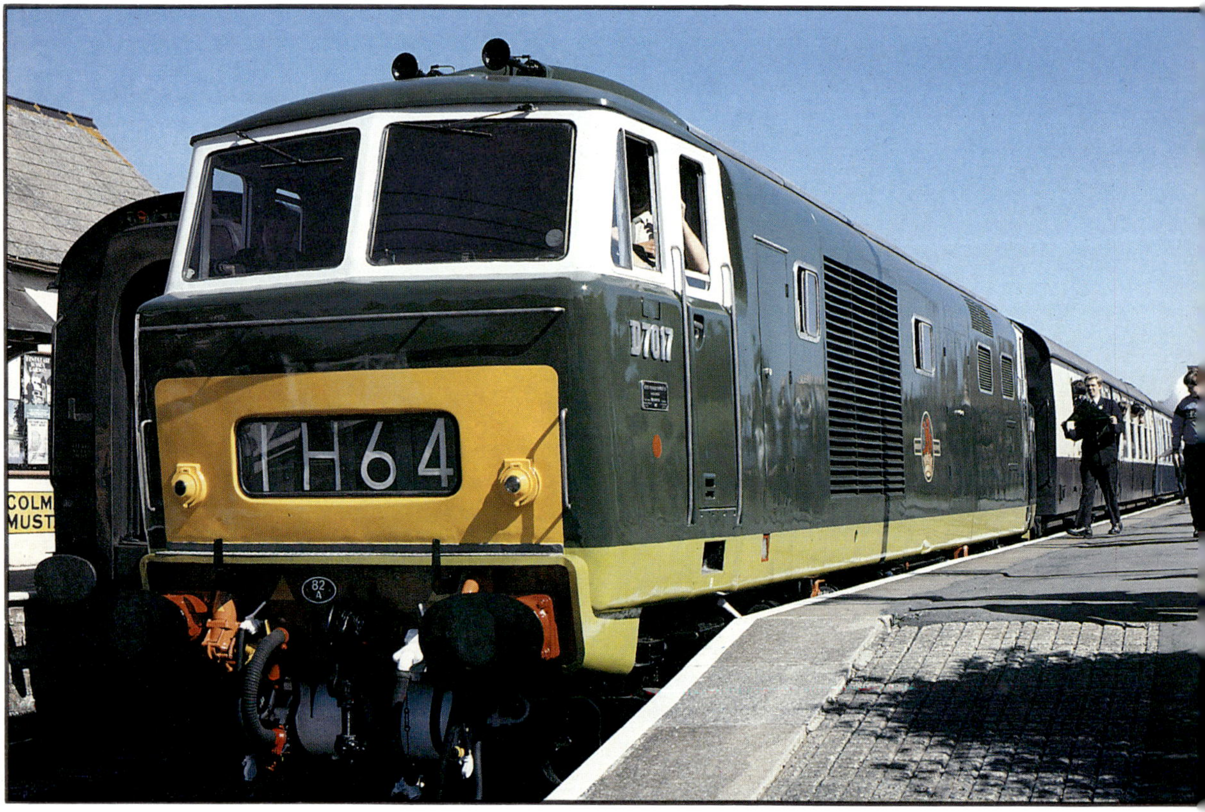
D7017
1H64

Previous Page Above:
A smartly turned out Rolls Royce diesel-engined Sentinel shunter stands in Bishops Lydeard Station at the head of a permanent way train. This useful loco was introduced to the railway in 1985.

Previous Page Below:
Approaching Crowcombe, a diesel Class 14 and pannier tank No. 6412 work together with a train destined to carry children on a visit to 'Santa'.

Above:
Class 14 diesel-hydraulic locos Nos. D9526 and D9551, seen here working a passenger train through Crowcombe. Occasionally these locos substitute for some DMU services.

Below:
Ruston & Hornsby-built Class 07 No. D2994 here in charge of a permanent way train, takes on fuel at Minehead's bay platform.

Pannier tank loco No. 6412 re-wheeling at the Minehead shed in September, 1987. Two road cranes make ready to lift the engine.

Above:
Carefully aligned No. 6412 is gently lowered into position to complete re-wheeling.

Left:
GWR pannier tank No. 6412 simmers between workings at Minehead Station in April 1986. Industrial loco JENNIFER, built by Hudswell Clarke & Co., now undergoing restoration, is in the background.

The extraordinary length of Minehead Station platform is evident from the water tower as No. 6412 heads a diesel multiple unit (DMU) away to Bishops Lydeard. Later the same engine will haul the QUANTOCK BELLE dining train from Bishops Lydeard back to Minehead.

Above:
Almost 26 years separate these two fine products of Swindon works. I have a good fire going in EVENING STAR, the coaling is being attended to and it is time for breakfast!

Left:
Prairie tank loco No. 5572 on the road from Keighley and Worth Valley Railway to Bishops Lydeard on 14th April, 1987 — the second of its two visiting seasons to the WSR. The engine is based at the Great Western Society's Didcot Centre.

Right:
Prairie tank loco No. 5572 in Blue Anchor Station while working a local train from Minehead. The coaches are about to be transferred to the down platform.

Below:
Evening 'on shed' at Minehead with pannier tank No. 6412, prairie tank No. 5572, and Bagnall saddle tank, VICTOR, in the background, on the coaling dock.

Left:
Prairie tanks Nos. 4561 and 5542 suffered the fate of years in Barry Scrapyard. Both had to be professionally stripped of claddings at Bishops Lydeard after arrival in September 1975.

Below:
After 14 years in scrapyard condition; rebuilding is now well under way — at the back of Minehead shed.

The moment of truth. Late at night on 15th August, 1989, on Minehead shed, prairie tank No. 4561 has a fire in for the first time.

Chapter 4

ROLLING STOCK

92220

Left:
A typical Sunday dining car train in the early 1980s steams through Washford and passes the Somerset & Dorset Trust's sidings.

Below:
The MINEHEAD PULLMAN, predecessor of the QUANTOCK BELLE dining car train, clean and smart with VULCAN in charge, waiting at Bishops Lydeard. Behind the loco is a second class non-corridor coach. The Bagnall saddle tank locos would be hard pushed to work a six-coach train, including the Pullmans, over the WSR gradients, and would often slip wildly — in adverse conditions —on the climb to Crowcombe.

Right:
The present QUANTOCK BELLE full dining service train in operation with variable motive power. Prairie tank No. 5572, is seen here at Bishops Lydeard.

Below:
Pannier tank No. 6412 in charge of a special charter working for which the entire train has been privately booked. The QUANTOCK BELLE is very popular with weekend diners and is booked well ahead.

Left:
GWR Sleeping Coach body No. 9038. How this 1897 Swindon-built first class coach came to the WSR is a story on its own. Upon withdrawal from service in 1934 — one of three originally built — it was sold as living accommodation to a buyer in the village of Burton, near Stogursey in Somerset. It was then incorporated into a bungalow, thus remaining well protected from the elements.

Right and Below:
The year is 1986. With the refurbishment of the bungalow under way, the admirably preserved coach is donated to the WSR, provided it can be speedily removed to a convenient field nearby.

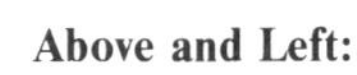

Above and Left:
Rails, old boiler tubes, much manpower, and a Field Marshal tractor with winch, complete the operation.

Right:
'Dragged through a hedge backwards'—No. 9038, a tribute to the Swindon builders, awaiting transport to the WSR, and restoration at Bishops Lydeard.

Partly restored ex-GWR coach, previously used as volunteer accommodation at Minehead, seen here with GW 'Toad' brake van and bogie sets in Bishops Lydeard siding.

Above:
Painted in an early livery of the railway — maroon and cream — to be seen here at Bishops Lydeard is an unusual 3-car Park Royal DMU set. As there are two motor brake second class cars in the make-up, this would give available support by two extra engines for the gradients, as well as extra seating accommodation, often required on summer Saturday services.

Left:
Interior of a Park Royal DMU set. On this occasion the Saturday service is well patronised, and passengers can enjoy excellent views from the large windows.

Above:
By 1982 one 2-car Park Royal DMU set had been purchased by the Diesel and Electric Group, and later repainted dark green with yellow stripe. Here it is coupled to the Craven set in Gala Week 1987, and working a Bishops Lydeard - Williton service.

Left:
During 1982 a Craven 2-car DMU set was purchased from BR (Eastern Region). After delivery by rail to Bridgwater Station sidings, it was transferred to Bishops Lydeard by road. Repainted green with yellow ends, the unit is seen here at Crowcombe.

Built in 1957 by the Gloucester Railway Carriage and Wagon Company, this 2-car DMU set is one of two delivered to the WSR in 1986 from Swanage. Repainted brown and cream with yellow ends, it is seen here in off-season service at Crowcombe.

Right:
For a short time after 1979 when second class, non-corridor, coach No. 46141 was at work on the railway, it would be seen supplementing Mark Is or Pullman trains. The problem with this ex-Kings Cross suburban coach was ticket checking on an otherwise full corridor train. Eventually it left the WSR for Wallingford.

Below:
There are times when stock, made redundant by BR, does not arrive on the WSR as quickly as it should and suffers the attentions of vandals. This Mark 1 coach arrived from Taunton with few windows intact.

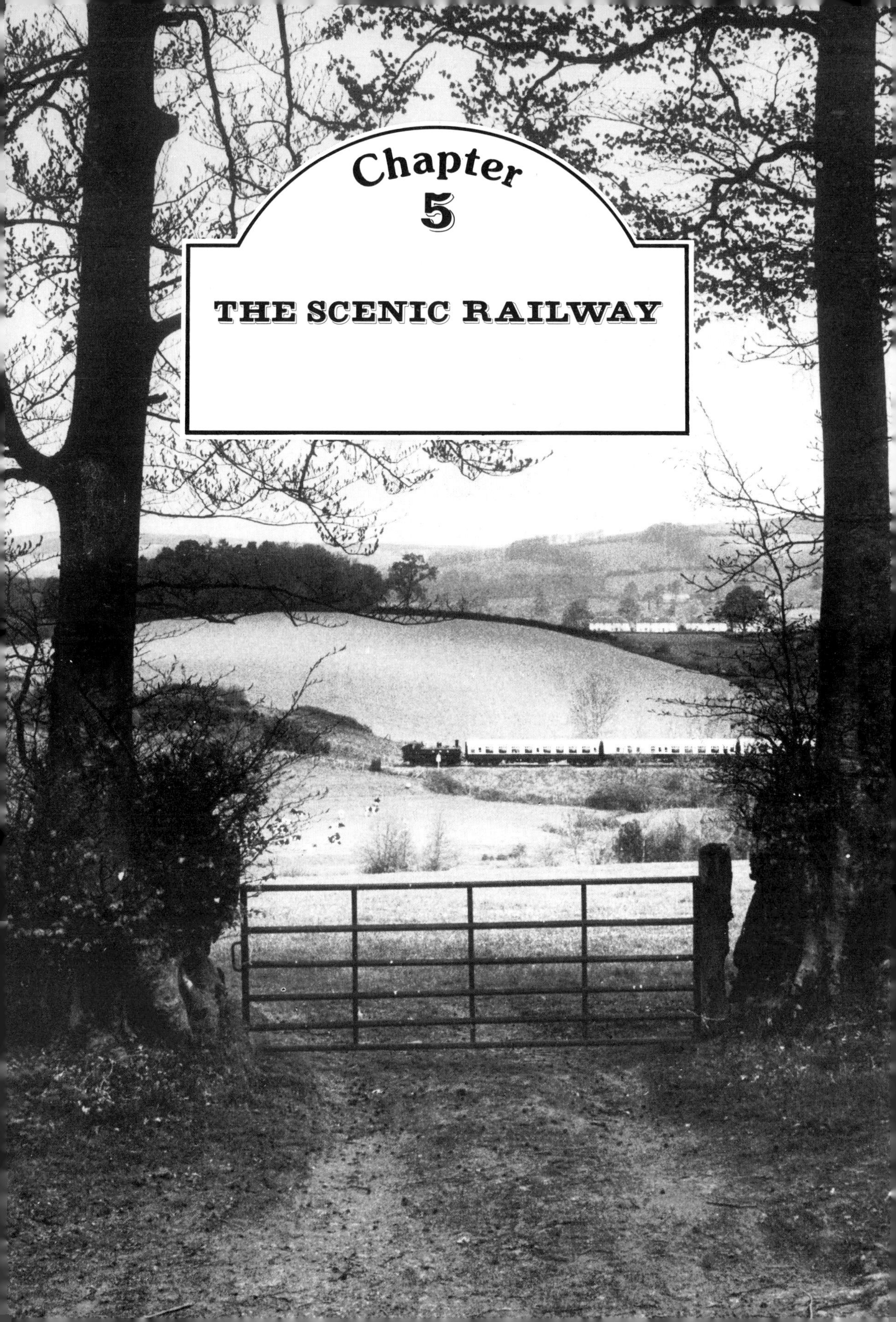

Chapter 5

THE SCENIC RAILWAY

Left:
Collet No. 3205 drifts over Ker Moor and into Blue Anchor through the heat haze of a July afternoon. Conygar Tower above Dunster in the background.

Below:
The still of a summer evening in the vale below the Brendon Hills is disturbed by No. 3205 working hard on Washford Bank.

Above:
The tide is out in Blue Anchor Bay as EVENING STAR and the QUANTOCK BELLE dining car set pass, in autumn sunshine, towards Blue Anchor Station.

Below:
A glimpse of industry as 7F No. 53808 rounds the curve past Wansbrough Paper Mill at Watchet. The Mill used to be served by rail.

Overleaf Above:
The beginning of 1988's season finds Collett designed No. 3205, in sparkling condition, heading an afternoon, Bishops Lydeard bound, train away from Minehead.

Overleaf Below:
Prairie tank No. 4561 returns to service in Gala Week, 1989, with a freight train from Minehead to Williton, here seen on Washford Bank.

From the signal box. Prairie tank No. 5572, on loan from the Didcot Railway Centre, crosses Ker Moor and 'opens up' for Blue Anchor.

Pannier tank No. 6412 hauls its train along the cliff top at Helwell Bay, Doniford. The sea has encroached uncomfortably near the line at this point, involving extensive defence works to protect the cliff face.

7F No. 53808, hauling a well filled train on August Bank Holiday from Williton in brilliant afternoon sunshine towards the Quantock Hills.

Overleaf Above:
'Bristol & Exeter' (pre-GWR) architecture in evidence on Williton's signal box and station building as EVENING STAR starts away with a train for Bishops Lydeard. In earlier BR days, double track extended over the river bridge (foreground).

Overleaf below:
Golden evening sunshine bathes the Quantock Hills near Woolston Moor as prairie tank No. 4561 slips past with a train for Minehead.

Previous Page Above:
On a fine summer morning beside the Quantocks, prairie tank No. 5572 steams past Leigh Bridge, on the long climb to Crowcombe.

Above:
Collett No. 3205 steams past Castle Hill, Williton, on the long climb to Crowcombe, some five miles distant.

Previous Page below:
The harvest is yet to be gathered in as pannier tank No. 6412, with steam in reserve, slips by the Quantock Hills near Bishops Lydeard.

Below:
Prairie tank No. 4561, newly restored, hauling a train near Bicknoller on its way to Minehead. Quantocks in the background.

Right:
Bagnall saddle tank power in the Quantocks, passing the site of Leigh Bridge loop and signal box.

Below:
'Ten on' and equal to the task. A wet Bank Holiday evening finds 7F No. 53808 near Leigh Woods crossing between Stogumber and Crowcombe.

Above:
The Quantock Hills at Crowcombe echo to the sound of the Diesel and Electric Group's Hymek (BR Class 35) No. D7017 on one of its rare appearances in service.

Left:
A bright clear May morning on the edge of the Quantock Hills finds pannier tank No. 6412 with an early season train for Crowcombe and Bishops Lydeard.

Above:
7F No. 53808 with a smart 'chocolate and cream' set of coaches makes the final effort into Crowcombe at the summit of the line.

Right:
A brief stop for 7F No.53808 among the trees at Crowcombe before it descends the final four miles of hidden country into Bishops Lydeard Station.

Above:
Collett No. 3205 makes good progress with a seven-coach train from Bishops Lydeard through the vale below the Quantock Hills in the neighbourhood of Combe Florey.

Below:
Bagnall saddle tank locos, VICTOR and VULCAN, tackle the climb to Crowcombe past Combe Florey in early preservation years.

A rare view of the track south east of Bishops Lydeard seen only by travellers in Gala Weeks and on special charter services. Collett No. 3205 is here seen approaching the main BR line at Norton Fitzwarren.

EVENTS

53808
2996

Left:
Space is at a premium at Bishops Lydeard Station when a vintage steam and transport rally is in progress. In this scene, pannier tank No. 6412 has arrived with visitors from stations down the line from Minehead.

Below:
By 1989 much more space was needed at Bishops Lydeard for the rally. So arrangements had to be made to borrow an adjoining field. Here are two veterans about to take on coal.

Traction engines at the Bishops Lydeard rally.

Above:
Annual Gala Weeks in September, with a special timetable, holiday attractions and demonstration freight trains operating between Minehead and Williton, are a feature of the WSR. The scene here is at Williton with pannier tank No. 6412 in charge.

Right:
Prairie tank No. 5572, with a Gala Week freight train from Minehead, waits in Blue Anchor Station for the arrival of a passenger train over the single track section from Williton

Above and Left:
Every year, during the summer, the paddle steamer WAVERLEY (above) and the BALMORAL (left) offer cruises in the Bristol Channel, in conjunction with steam trips on the WSR, and may often be seen tied up in Minehead harbour.

Right:
Since the re-opening of the line in 1976, various television companies have 'shot' the WSR for its own sake and as background for film subjects. Here, at Minehead, we see Bernard Goodsall filming 'The Seaside Trains' early in 1988 for the BBC.

Below:
In 1988 the BBC film crew was with us again, this time at Crowcombe, shooting scenes for 'The Lion, the Witch, and the Wardrobe', when the station assumed a 1940s look. Blue Anchor too was the location for a LWT production of an Agatha Christie Poirot story.

Above Left:
August 1980. The Lucas Aerospace Workers' experimental Road-Rail vehicle, under the watchful eye of John Doyle and the Westward Television camera crew, prepares to depart from Bishops Lydeard Station.

Above Right:
On the track and going well after climbing to the summit of the line at Crowcombe, the Road-Rail bus heads for Roebuck level crossing where it will take to the road again.

Below:
Limited storage capacity for water for engines working the line at peak times has meant calling in the fire brigade for a 'top-up'. On this occasion VICTOR was the recipient at Williton Station level crossing.

14th May, 1988. An immaculate Collett No. 3205 ready to double-head the 'Pines Express' with 7F No. 53808 from Minehead to Bishops Lydeard. This train was a 'special' for Somerset and Dorset Railway enthusiasts, whose museum is housed on the West Somerset Railway at Washford; 7F No. 53808 is owned by the Somerset and Dorset Railway Trust and leased to the WSR.

Probably the best known of all Somerset & Dorset railwaymen, Donald Beale, receives a superbly iced cake, presented to him by S & D staff at their re-union at Washford Station in August 1989. EVENING STAR stands in the sidings to the rear.

27th November 1988 finds 7F No. 53808 down among the cider barrels in Taunton Cider's works at Norton Fitzwarren. This was the furthest point any WSR train had reached, close beside the BR's West of England main line.

Chapter 7

EVENING STAR

The world-famous engine EVENING STAR belongs to the National Collection of Steam Locomotives and is normally housed at the National Railway Museum, at York. In 1989 it was loaned to the WSR for the season.

Bank Holiday blue skies over Minehead make a perfect setting for EVENING STAR heading away with a crowded afternoon train.

Right:
Following a long haul lasting several days by road from the National Railway Museum 9F No. 92220 EVENING STAR and tender, each on a low-loader, take the Cross Keys road junction at Norton Fitzwarren. Just a few miles to go . . .

Below:
Safely delivered to Bishops Lydeard. Time to unload and, following erection of ramps, engine and tender are eased, inch by inch, from their transporters. By the end of the afternoon, EVENING STAR, 9F No. 92220 and tender had been re-united on WSR metals and were then guarded overnight in the goods shed at Bishops Lydeard.

Above:
Morning on Minehead shed. The 9F is carefully prepared for a long day in service. Trains are fully booked, and all eyes will be focussed on the performance of EVENING STAR.

Below:
In the early morning at Minehead Station, the power of a giant remains to be awakened, as it rests beside pannier tank No. 6412, another very different product of Swindon's loco builders.

Overleaf Above:
EVENING STAR, a popular visitor to the West Somerset Railway in 1989, simmers on Washford sidings at the end of a Somerset and Dorset open day.

Overleaf Below:
In quieter mood and, with just a wisp of steam, EVENING STAR slides past Kingswood on the down gradient to Williton.

Previous Page Above:
Crowcombe's well-tended station, basking in spring afternoon sunshine, welcomes the arrival of EVENING STAR.

Previous Page Below:
Summer time at Combe Florey with EVENING STAR at the head of a passenger train in the depths of the West Somerset countryside.

May 1989 — passengers crowd on to Bishops Lydeard Station platform to see and travel behind EVENING STAR. The season was to break records for numbers of passengers carried on the railway.

Left:
'Park and Ride' seems to be the order of the day on Spring Bank Holiday Sunday 1989 at Bishops Lydeard Station as EVENING STAR prepares to depart with the 4.10 p.m. train to Minehead.

Below:
The height of summer. The 9F, in charge of the QUANTOCK BELLE dining car train, makes a brief stop at Crowcombe.

Above:
Castle Hill, Williton. A favourite spot to set up your camera, especially when EVENING STAR steams past with the QUANTOCK BELLE on a sunny afternoon.

Right:
Summer sunshine casts shadows across Minehead shed before there is much activity around 7F No. 53808 and EVENING STAR, mainstays of the summer service.

A good supply of 'Brasso' and elbow grease is called for when faced with the generous array of pipework on EVENING STAR. Lighting-up wood is on the footplate for next day, while pannier tank No. 6412 is showing a chimney on the coaling dock.

With North Hill, Minehead, in the background, 9F EVENING STAR is about to leave for Bishops Lydeard.